To my mom and dad, with love
—E.M.

The author and editor would like to thank Thomas R. Holtz Jr., PhD, Principal Lecturer in Vertebrate Paleontology, Department of Geology, University of Maryland, College Park, for assistance in the preparation of this book.

Random House Books for Young Readers
An imprint of Random House Children's Books
A division of Penguin Random House LLC
1745 Broadway, New York, NY 10019
penguinrandomhouse.com
rhcbooks.com

Library of Congress Cataloging-in-Publication Data is available upon request.
ISBN 978-0-593-37728-4 (trade) — ISBN 978-0-593-37729-1 (lib. bdg.)
ISBN 978-0-593-37730-7 (ebook)

Manufactured in China
10 9 8 7 6 5 4 3 2 1

The authorized representative in the EU for product safety and compliance is Penguin Random House Ireland, Morrison Chambers, 32 Nassau Street, Dublin D02 YH68, Ireland, https://eu-contact.penguin.ie.

DINOSAUR DAY

by Edward Miller

Beginner Books®
A Division of Random House

Early each morning
the dinosaurs rise.
The sun comes up.
They open their eyes.

Everyone looks
for something to eat.

Plant-eaters want plants.

Meat-eaters want meat.

Some are big.
Some are small.
This one here
is very tall!

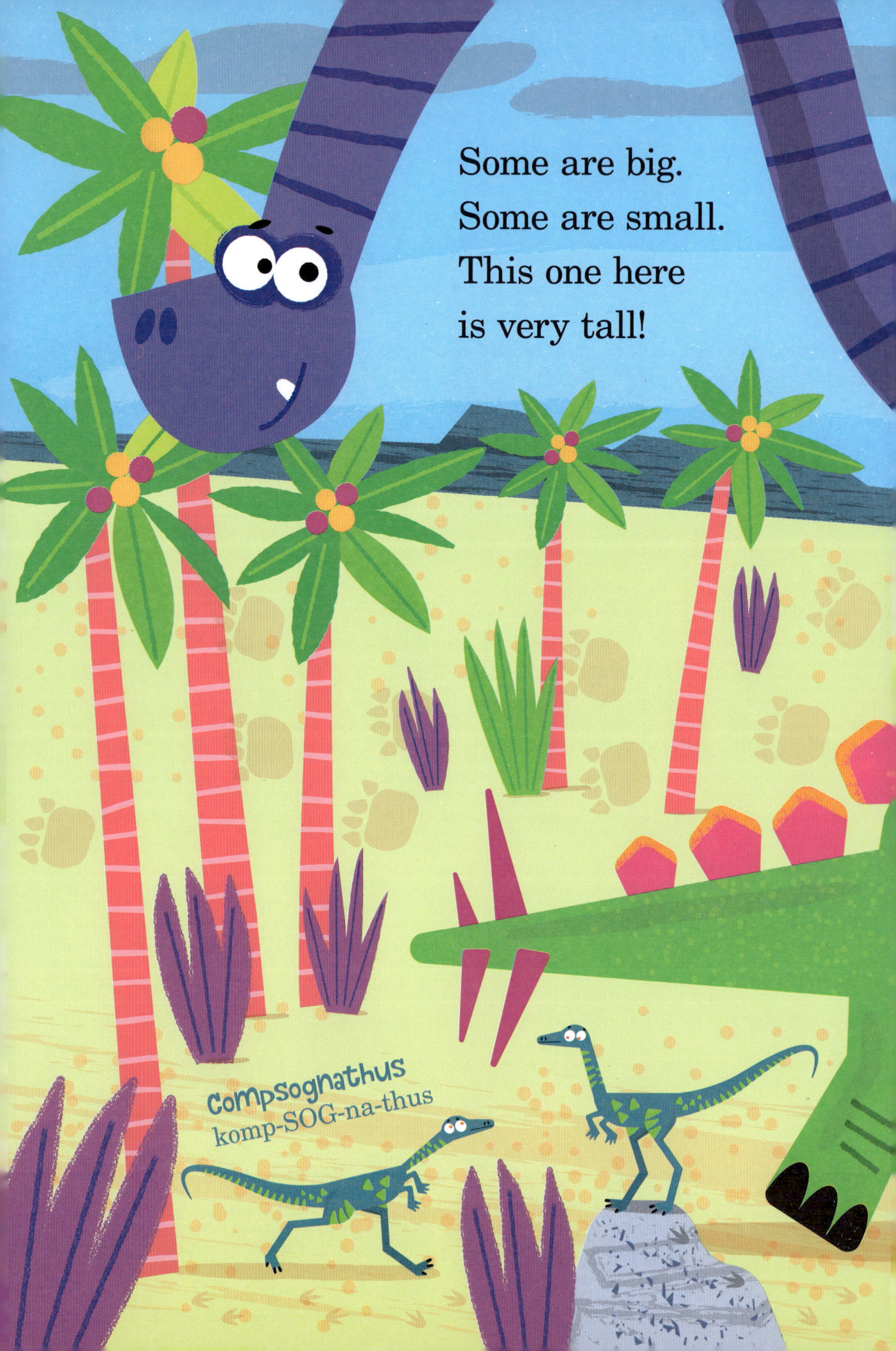

Sauroposeidon
SAWR-o-po-SYE-don
Stegosaurus
STEG-o-SAWR-us

Some have feathers.

Some have scales.

Carnotaurus
KAHR-no-TAWR-us

Some have spikes
along their tails!

Taking a dip
in the salty sea,
this dino eats fish.
One! Two! Three!

Spinosaurus
SPY-no-SAWR-us

Brachiosaurus
BRAK-ee-o-SAWR-us

Another dino
in a swamp
feeds on leaves.
Chomp!
Chomp!
Chomp!

Flying reptiles
in the sky.
They flap their wings.
Hello! Goodbye!

Pterodactylus antiquus
tair-o-DAK-til-us an-TEEK-wus
Quetzalcoatlus
KET-sal-ko-AHT-lus

Swimming reptiles
in a lake.
They splish and splash—
then the earth starts to shake!

Plesiosaurus
PLEE-see-o-SAWR-us

Edmontosaurus
ed-MON-toe-SAWR-us
Triceratops
tri-SAIR-a-tops

A volcano rumbles.
It is going to blow!
Quick, dinos, run
from the hot lava flow!

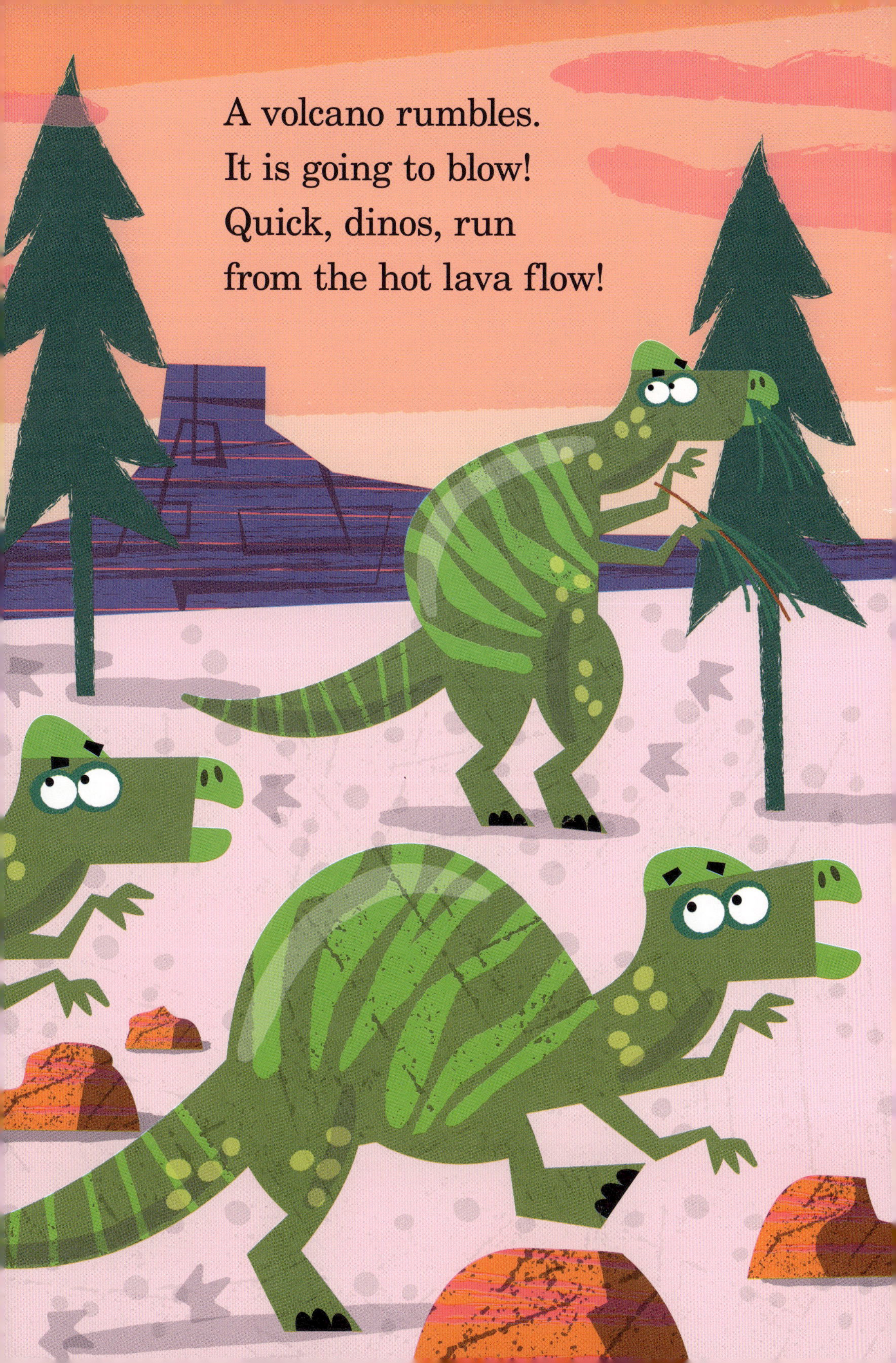

They run, then stop
to take a break.

But have they made
a big mistake?

There is another
dino near.
A dino that
fills them with fear!

T. rex walks
on long, strong feet.
She's looking for
some meat to eat.

She likes to hunt.
She does it well.
She has a super
sense of smell!

But when she runs
after her prey,
she trips and falls!
They get away!

With a boo-boo
on her knee,
T. rex goes home
to her family.

Because to her mama
and papa, too,
she is ALWAYS their baby.
Just like YOU!

When the sun goes down,
it is time for bed.

Time each dozy dino
rests their head.

Some sleep curled up
like birds of a feather.

Some sleep in groups.
They are safer together.

Do dinosaurs dream?
We cannot say.
Maybe of food?
Or friends? Or play?

Or maybe a perfect
dinosaur day?

If YOU were a dinosaur,
what would YOU say?